solar system

URANUS,

NEPTUNE,

AND THE DWARF PLANETS

Rosalind Mist

QEB

QEB Publishing

Words in bold can be found in the glossary on page 22.

Library of Congress Control Number:
2008012595

ISBN 978 1 59566 665 9

Printed and bound in China

Author Rosalind Mist
Consultant Terry Jennings
Editor Amanda Askew
Designer Melissa Alaverdy
Picture Researcher Maria Joannou

Publisher Steve Evans
Creative Director Zeta Davies

Picture credits
(fc=front cover, t=top, b=bottom, l=left,
r=right, c=centre)

Calvin J Hamilton 9tcr

Corbis Bettmann 9b, Denis Scott 21t, Firefly
Productions 8–9, Stapleton Collection 8t

NASA fc, 1t, NASA JPL 1b, 4b, 5, 6–7, 9tcl,
9tr, 10t, European Space Agency 11,14, 15t,
18t, 20t, 21c, 21b, 23, Dr R Albrecht/ESA/ESO
Space Telescope European Coordinating
Facility 18b, Johns Hopkins University
Applied Physics Laboratory/
Southwest Research Institute
18–19, 24, JPL 2–3, 9Tl, 9Tc,
10b, 12t, 13, 15b, JPL/STScl 7r,
JPL-Caltech 20b

Science Photo Library
Mark Garlick 16–17

Shutterstock 4T, 4–5

Contents

The Solar System   4

Uranus   6

Discovery   8

Voyager missions   10

Neptune   12

Rings and moons   14

The dwarf planets   16

Pluto   18

Eris and Ceres   20

Glossary   22

Index   23

Notes for parents and teachers   24

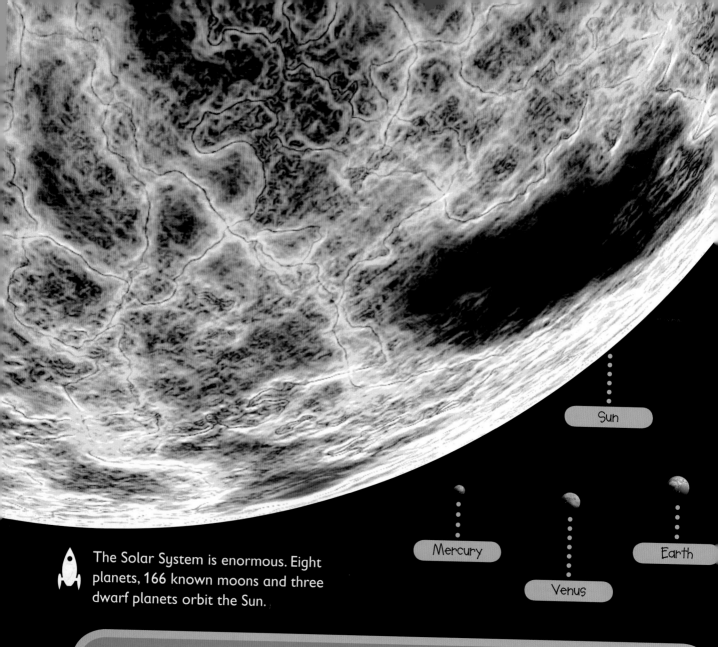

Sun

Mercury

Venus

Earth

 The Solar System is enormous. Eight planets, 166 known moons and three dwarf planets orbit the Sun.

The Solar System is held together by an invisible force called **gravity**. On Earth, gravity stops people from floating into space!

The **International Space Station** takes 90 minutes to go once around the Earth. The **astronauts** that live there are falling freely around the Earth and feel weightless.

The Solar System

The Solar System is made up of the Sun, and everything that orbits, or circles, it.
This includes the planets and their moons, as well as **meteors**, **asteroids**, and **comets**.

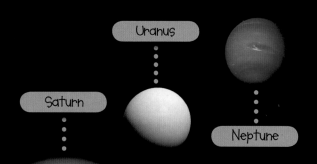

Uranus

Saturn

Neptune

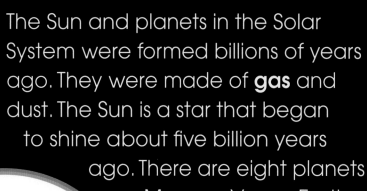

Mars

Jupiter

The Sun and planets in the Solar System were formed billions of years ago. They were made of **gas** and dust. The Sun is a star that began to shine about five billion years ago. There are eight planets —Mercury, Venus, Earth, Mars, Jupiter, Saturn, Uranus, and Neptune.

STAR FACT!
Earth seems very big to us, but Jupiter is 11 times wider than Earth and the Sun is ten times wider than Jupiter!

The sizes of the planets are roughly to scale, but the distances between them are not to scale.

Uranus

Uranus is the third largest planet, after Jupiter and Saturn. About four Earths could fit across it. Uranus comes after Saturn, and is the seventh planet from the Sun. It takes 84 years to orbit the Sun.

Uranus is a huge blue-green ball of different gases and it has a core, or middle, of rock and ice. Its surface looks smooth, but it is not solid, so **spacecraft** cannot land on it.

STAR FACT!
On Uranus, winter lasts for 21 years and half the planet is in darkness. During summer, which also lasts for 21 years, the Sun shines all the time.

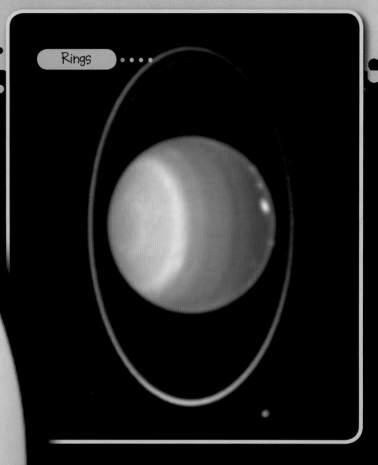

Rings

Uranus' rings are very faint and difficult to see. Scientists have to use special **telescopes** to see the rings.

The rings of Uranus go from top to bottom. Scientists think that many years ago, Uranus was hit by something very large. It was knocked over and now it spins on its side.

Uranus is made from gases. A gas called **methane** makes it look blue-green in color.

Discovery

Five planets have been known about for thousands of years because people could see them moving slowly in the sky. Uranus was the first new planet to be discovered. On 13 March 1781, **astronomer** William Herschel saw something using his telescope that was not on his **star chart**.

Dome

Telescope

As well as discovering Uranus, Herschel also measured the heights of mountains on Earth's Moon and discovered four of Uranus' moons.

This telescope in Spain is named after William Herschel.

STAR FACT!
Uranus can sometimes be seen from Earth. It looks like a very faint star.

Miranda   Ariel   Umbriel   Titania   Oberon

Astronomers realized that it was a planet, twice as far away from the Sun as Saturn. They also worked out that there should be another planet as well, because Uranus was never quite where they thought it should be. It was being tugged by the gravity of something large. They searched with telescopes and found Neptune in 1846.

 Uranus has at least 27 moons. These are the five main moons. Herschel found Titania and Oberon. Ariel and Umbriel were discovered by William Lassell in 1851. Miranda was not seen until 1948.

 William Herschel's largest telescope took two years to make and was the largest telescope in the world for more than 50 years.

Voyager missions

In 1977, two spacecraft called *Voyager 1* and *Voyager 2* were sent into the Solar System. They both visited Jupiter and Saturn. *Voyager 2* went on to explore Uranus and Neptune.

Voyager 2 took nine years to reach Uranus. It took pictures of the planet and its five main moons. It also discovered ten new moons. Three years later, *Voyager 2* flew over Neptune's north pole and past Neptune's largest moon, Triton.

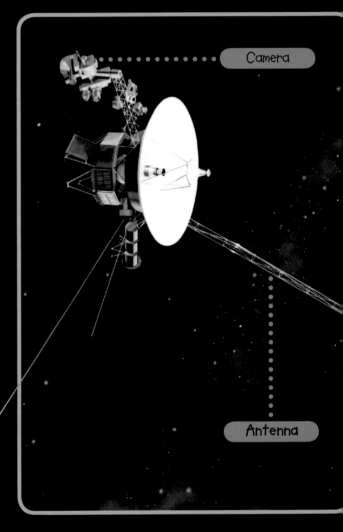

Camera

Antenna

Voyager 2 has a camera so scientists can find out more information about the planets.

The south pole of Uranus' Moon Miranda was photographed by *Voyager 2* as it flew by. The moon is only 300 miles across.

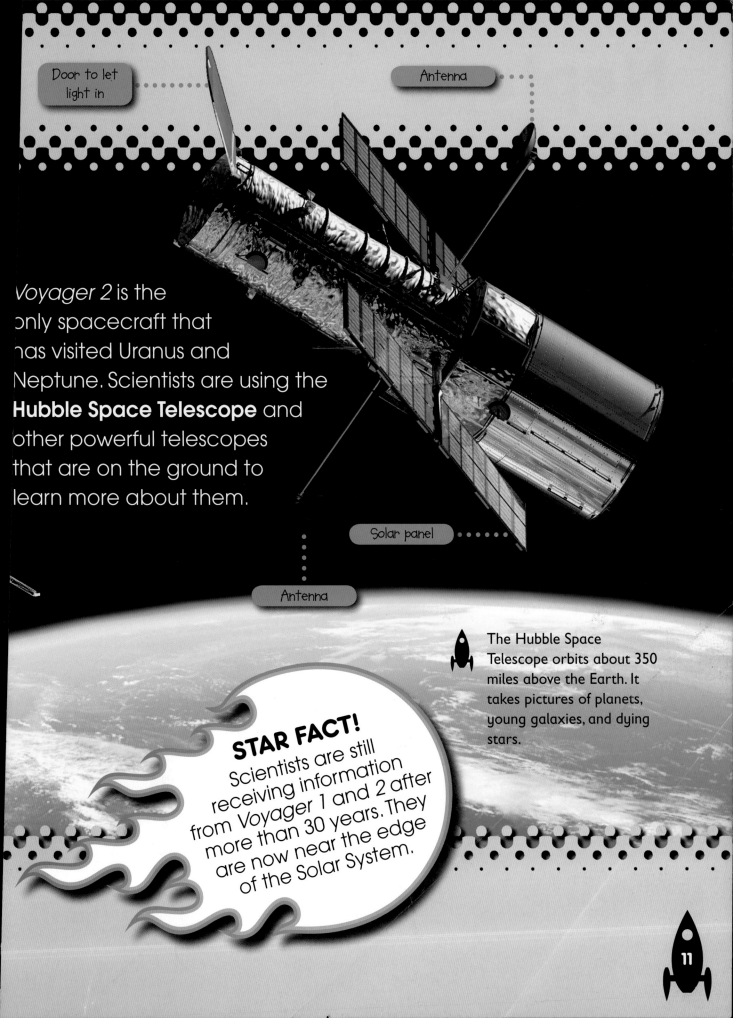

Door to let light in

Antenna

Voyager 2 is the only spacecraft that has visited Uranus and Neptune. Scientists are using the **Hubble Space Telescope** and other powerful telescopes that are on the ground to learn more about them.

Solar panel

Antenna

The Hubble Space Telescope orbits about 350 miles above the Earth. It takes pictures of planets, young galaxies, and dying stars.

STAR FACT!
Scientists are still receiving information from Voyager 1 and 2 after more than 30 years. They are now near the edge of the Solar System.

Neptune

Neptune is the furthest planet in the Solar System.

Neptune is so far away that it takes 165 years to go around the Sun once. Beyond Neptune, there are only dwarf planets, comets, and asteroids.

Neptune is the fourth largest planet in the Solar System. It is slightly smaller than Uranus and 58 Earths could fit inside it.

Clouds

Neptune looks mainly blue, but there are some white clouds. These bands of cloud are up to 120 miles wide.

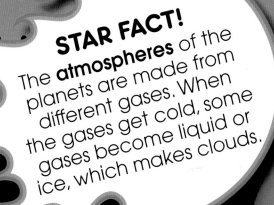

STAR FACT!
The **atmospheres** of the planets are made from different gases. When the gases get cold, some gases become liquid or ice, which makes clouds.

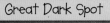
Great Dark Spot

 Neptune is made from different gases. One of the gases, called methane, makes the planet blue in color.

Neptune is the windiest planet in the Solar System. Winds can reach speeds of more than 1000 miles an hour —seven times faster than the strongest storms on Earth. There are huge storms that are about the size of the Earth.

 The Great Dark Spot was discovered by *Voyager 2* in 1989. It has now disappeared.

Where do clouds come from?
. . . .

Fill a glass with icy water and leave it for 30 minutes. Look at the outside of the glass. What do you see? When warm air hits the cold glass, water droplets in the air turn into liquid water! Clouds are formed in a similar way to this, when warm air hits cold air.

Rings and moons

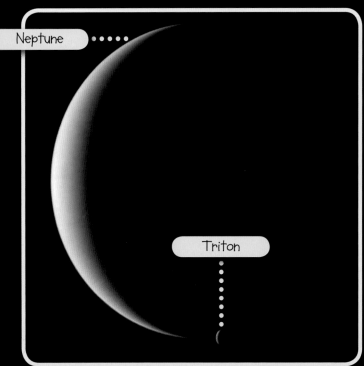

Neptune

Triton

Triton goes around Neptune in the opposite direction to normal moons. It is 1600 miles across —a little bigger than Pluto.

Both Uranus and Neptune have rings and lots of moons.
So far, astronomers have found 27 moons orbiting Uranus. Neptune has 13 moons.

Neptune's largest moon is Triton. It was discovered soon after astronomers discovered the planet. Triton is rocky, icy, and very cold —the coldest known object in the Solar System. It is red-brown and blue-green in color with pink ice caps.

Ice
Triton is covered in water ice and other frozen gases. Draw a line halfway up a small plastic cup. Fill it with water up to the line. Now freeze the water. What happens to it?

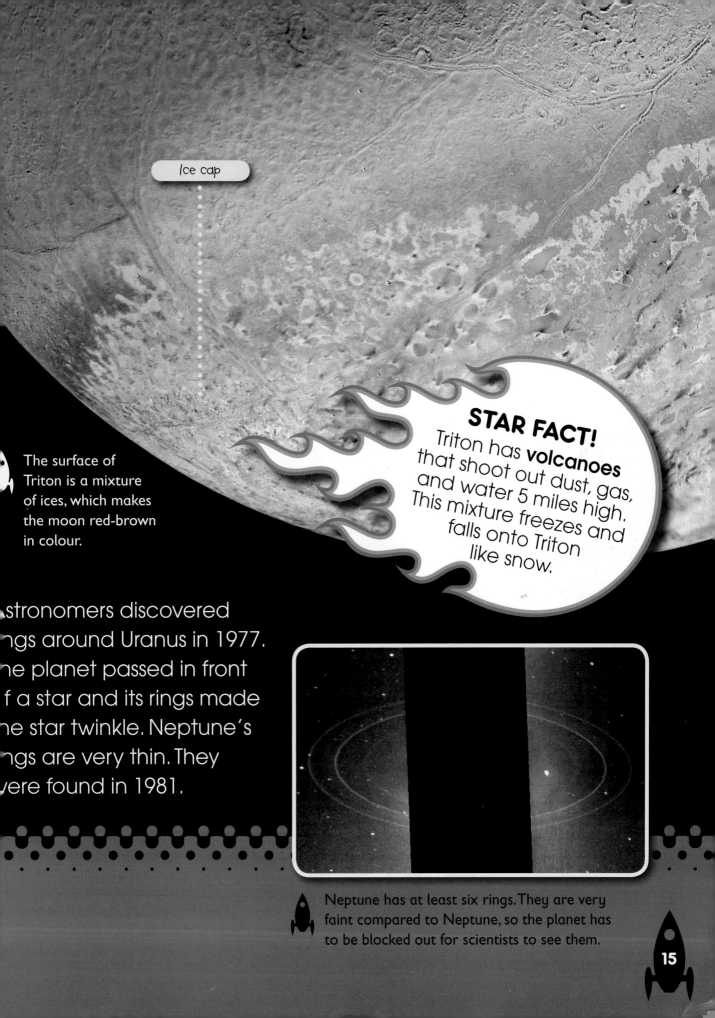

Ice cap

The surface of Triton is a mixture of ices, which makes the moon red-brown in colour.

STAR FACT!
Triton has **volcanoes** that shoot out dust, gas, and water 5 miles high. This mixture freezes and falls onto Triton like snow.

stronomers discovered ngs around Uranus in 1977. ne planet passed in front f a star and its rings made ne star twinkle. Neptune's ngs are very thin. They vere found in 1981.

Neptune has at least six rings. They are very faint compared to Neptune, so the planet has to be blocked out for scientists to see them.

The dwarf planets

Astronomers have recently found more large objects moving around the Sun.
They had to decide if these objects were new planets. Planets are bodies, or objects, that move around the Sun.

Planets are round like a ball. They are big enough to have cleared any rock and ice in their way. These either hit the planet or they were moved out of the way.

 Eris and Pluto are further from the Sun than Neptune. Ceres is in the Asteroid Belt between Mars and Jupiter.

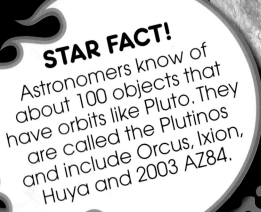

Eris

STAR FACT!
Astronomers know of about 100 objects that have orbits like Pluto. They are called the Plutinos and include Orcus, Ixion, Huya and 2003 AZ84.

Pluto

Ceres

Pluto was once called a planet. Although Pluto is round and moves around the Sun, it is not big enough to be a planet. Astronomers now call Pluto a dwarf planet. So far astronomers have found two other dwarf planets, Eris and Ceres. There are probably many more.

Make a planet

Dust a tray with a powder, such as flour or cocoa. Roll a ball of modeling clay across the tray. What happens? The planets grew bigger by collecting dust. Planets are so big that their gravity pulls dust towards them, too, so they collect more than just the bits in their way.

Pluto

Pluto was discovered in 1930 and was once the ninth planet. However, scientists now say it is a dwarf planet. Pluto is smaller than both the Moon and Neptune's moon, Triton. It is cold and icy, and brown in color.

Pluto has three moons. The biggest moon is called Charon, which was discovered in 1978. Charon is 800 miles across, which is more than half the size of Pluto. Pluto has two other small moons called Hydra and Nix. They were discovered in 2005.

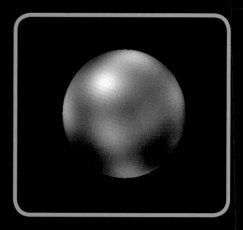

 Pluto is so far away that it just looks gray in color to us, even with powerful telescopes.

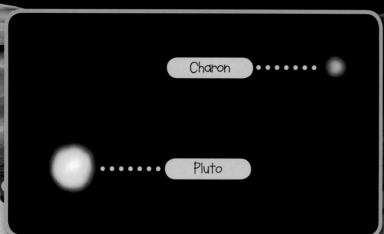

Charon •••••••

Pluto

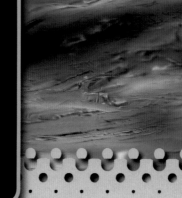 Pluto and Charon are so far away, they are difficult to see, even with large telescopes.

luto is so far away that it has not et been visited by spacecraft. here is a spacecraft called *New Horizons* on its way to Pluto. It was sent into space n 2006, but it will not get there until July 2015.

STAR FACT!
Pluto's name was suggested by an 11-year-old schoolgirl from Oxford.

New Horizons is the fastest spacecraft so far. It left Earth traveling at more than 10 miles a second and passed Jupiter after just one year.

Power supply

Antenna dish

Eris and Ceres

All the dwarf planets are smaller than the Earth's Moon.

Eris is the largest dwarf planet —it is slightly larger than Pluto. Ceres is only one-third of the size of Earth's Moon.

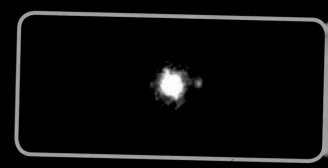

 Eris is so far away and faint, even the Hubble Space Telescope cannot see it clearly. Its moon Dysnomia is on the right.

Eris is one of the most distant large objects discovered so far in the Solar System. It is a very long way from the Sun—about three times further than Pluto. It is so far away that it is very faint and was only found in 2003. Eris has a moon called Dysnomia.

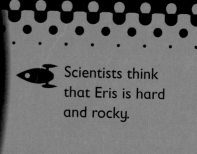 Scientists think that Eris is hard and rocky.

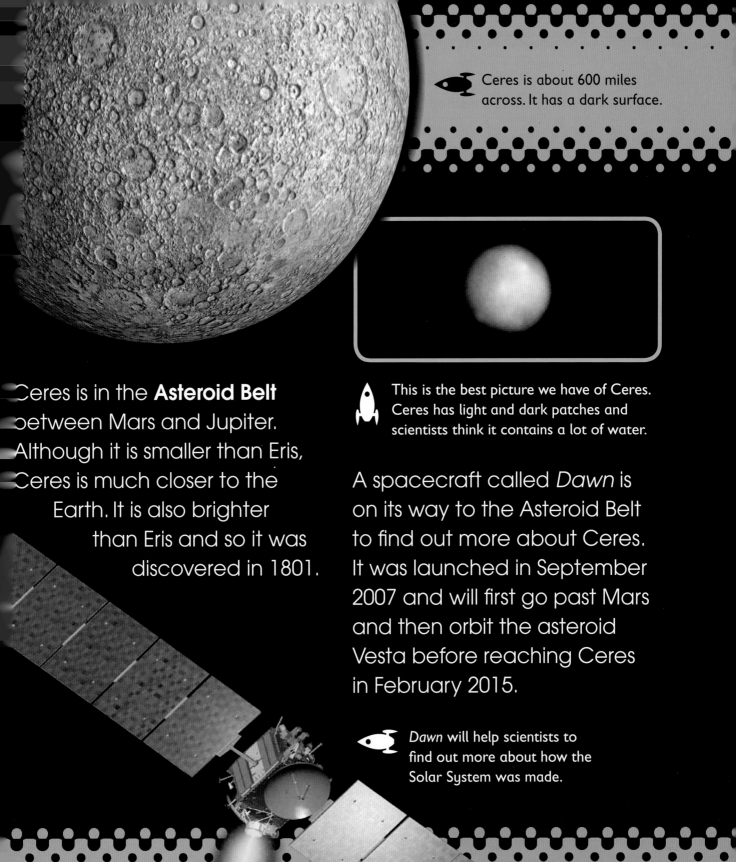

Ceres is about 600 miles across. It has a dark surface.

This is the best picture we have of Ceres. Ceres has light and dark patches and scientists think it contains a lot of water.

Ceres is in the **Asteroid Belt** between Mars and Jupiter. Although it is smaller than Eris, Ceres is much closer to the Earth. It is also brighter than Eris and so it was discovered in 1801.

A spacecraft called *Dawn* is on its way to the Asteroid Belt to find out more about Ceres. It was launched in September 2007 and will first go past Mars and then orbit the asteroid Vesta before reaching Ceres in February 2015.

Dawn will help scientists to find out more about how the Solar System was made.

Glossary

Antenna
A wire that is used for receiving radio and television signals.

Asteroid
A large lump of rock, too small to be a planet or dwarf planet.

Asteroid Belt
Area between Mars and Jupiter where there are lots of asteroids.

Astronaut
A person who travels in space.

Astronomer
A scientist who studies the Solar System, stars, and galaxies.

Atmosphere
A layer of gases around a planet or moon.

Comet
An object in space made of rock and ice.

Gas
A substance, such as air, that is not solid or liquid. Gas cannot usually be seen.

Gravity
Attractive pulling force between any massive objects.

Hubble Space Telescope
A telescope that is orbiting the Earth.

International Space Station
Large space laboratory where astronauts can live for months.

Meteor
A glowing trail in the sky left by a small piece of rock from space.

Methane
Natural gas.

Orbit
The path of one body around another, such as a planet around the Sun.

Spacecraft
A vehicle that travels in space.

Star chart
A map of the night sky showing names and positions of stars.

Telescope
An instrument that makes things in the distance seem large.

Volcano
A place where magma comes to the surface.

Index

Ariel 9
Asteroid Belt 16, 21
asteroids 5, 12

Ceres 16, 17, 20–21
Charon 18

dwarf planets 4, 12, 16–17,
    18, 20

Eris 16, 17, 20–21

Great Dark Spot 13

Herschel, William 8, 9
Hubble Space Telescope 11,
    20

Lassell, William 9

Miranda 9, 10
moons 4, 5, 8, 9, 10, 14, 18, 20

Neptune 5, 9, 10, 11, 12–13,
    14–15

Oberon 9

Pluto 16, 17, 18–19

rings 7, 14, 15

spacecraft 6, 10, 19, 21

Titania 9
  Triton 10, 14, 15

    Umbriel 9
    Uranus 5, 6–7, 8, 9, 10, 11, 14

    volcanoes 15
    Voyager spacecraft 10, 11,
      13

What are *Voyager 1* and *2* doing now? They have both now entered one of the furthest regions of the Solar System where particles from the Sun start to meet those from outer space. They are continuing to make measurements to help scientists understand the outermost parts of the Solar System.

Unlike most liquids, water actually expands when it cools and becomes ice. This is why pipes can burst if they freeze and why you should not put full bottles of water in the freezer—as the water expands, it can burst bottles and pipes!

Why does Pluto's Moon, Charon, seem to stay in the same place in the sky? We see the same side of the Moon because it is rotating at the same rate as its orbit around the Earth. The Earth spins independently so the Moon rises and sets. Pluto and Charon are "locked" together facing each other as they orbit, so from Pluto Charon seems to hover in the same place in the sky!

The discovery of Eris caused problems for astronomers, particularly when they realized it is larger than Pluto. If Pluto is a planet then Eris has to be, too. With the likelihood of many more similar bodies, it looked like there would be many more planets. However, some astronomers had long argued that Pluto was not a "proper" planet and so it was decided to define a new category— the dwarf planet.

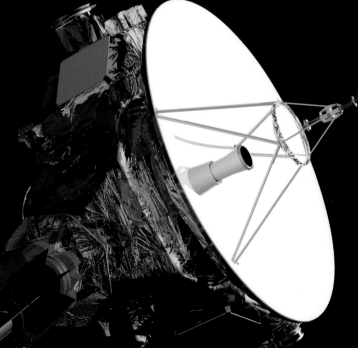